— *of* —

NEW JERSEY

Tamara Eder

with contributions from **Ian Sheldon**

© 2002 by Lone Pine Publishing
First printed in 2002   10 9 8 7 6 5 4 3 2 1
Printed in Canada

All rights reserved. No part of this work covered by the copyright hereon may be reproduced or used in any form or by any means—graphic, electronic or mechanical—without the prior written permission of the publisher, except for reviewers, who may quote brief passages. Any request for photocopying, recording, taping or storage in an information retrieval system of any part of this book shall be directed in writing to the publisher.

THE PUBLISHER: LONE PINE PUBLISHING

1901 Raymond Avenue SW, Suite C	10145-81 Avenue
Renton, WA 98055	Edmonton, AB T6E 1W9
USA	Canada

Lone Pine Publishing Website: http://www.lonepinepublishing.com

National Library of Canada Cataloguing in Publication Data

Eder, Tamara, (date)
   Animal tracks of New Jersey

   Includes bibliographical references and index.
   ISBN 1-55105-341-1

   1. Animal tracks—New Jersey—Identification.   I. Sheldon, Ian, (date) II. Title.
QL768.E363 2001   591.47'9   C2001-911012-X

Editorial Director: Nancy Foulds
Editor: Volker Bodegom
Proofreaders: Lee Craig, Genevieve Boyer
Production Coordinator: Jen Fafard
Design, Layout & Production: Volker Bodegom, Monica Triska
Cover Design: Elliot Engley
Technical Contributor: Mark Elbroch
Cartography: Volker Bodegom
Animal Illustrations: Gary Ross, Horst Krause, Ian Sheldon, Ewa Pluciennik
Track Illustrations: Ian Sheldon
Cover Illustration: Red Squirrel by Gary Ross
Scanning: Elite Lithographers Ltd.

We acknowledge the financial support of the Government of Canada through the Book Publishing Industry Development Program (BPIDP) for our publishing activities.

PC: P4

CONTENTS

Introduction 4
  How to Use This Book 5
  Tips on Tracking 6
  Map 7
  Terms & Measurements 9

Mammals 17
  White-tailed Deer 18
  Horse 20
  Black Bear 22
  Domestic Dog 24
  Coyote 26
  Red Fox 28
  Gray Fox 30
  Bobcat 32
  Domestic Cat 34
  Raccoon 36
  Opossum 38
  Harbor Seal 40
  River Otter 42
  Mink 44
  Long-tailed Weasel 46
  Short-tailed Weasel 48
  Striped Skunk 50
  Black-tailed Jackrabbit 52
  European Hare 54
  Eastern Cottontail 56
  Porcupine 58
  Nutria 60
  Beaver 62
  Muskrat 64
  Woodchuck 66
  Eastern Chipmunk 68
  Eastern Gray Squirrel 70
  Fox Squirrel 72
  Red Squirrel 74
  Southern Flying Squirrel 76
  Norway Rat 78
  Woodland Vole 80
  White-footed Mouse 82
  Meadow Jumping Mouse ..84
  Masked Shrew 86
  Star-nosed Mole 88

Birds, Amphibians & Reptiles 91
  Mallard 92
  Herring Gull 94
  Great Blue Heron 96
  Common Snipe 98
  Spotted Sandpiper 100
  Ruffed Grouse 102
  Great Horned Owl 104
  American Crow 106
  Northern Flicker 108
  Dark-eyed Junco 110
  Northern Cardinal 112
  Frogs 114
  Toads 116
  Salamanders & Newts 118
  Skinks 120
  Turtles 122
  Snakes 124

Appendix
  Track Patterns
    & Prints *126*
  Print Comparisons *135*
  Bibliography *138*
  Index *139*

INTRODUCTION

If you have ever spent time with an experienced tracker, or perhaps a veteran hunter, then you know just how much there is to learn about the subject of tracking and just how exciting the challenge of tracking animals can be. Maybe you think that tracking is no fun, because all you get to see is the animal's prints. What about the animal itself—is that not much more exciting? Well, for most of us who don't spend a great deal of time in the beautiful natural areas of New Jersey, the chances of seeing the elusive Red Fox or the fun-loving River Otter are slim. The closest that we may ever get to some animals may be through their tracks, which can inspire a very intimate experience. Remember, you are following in the footsteps of the unseen—animals that are in pursuit of prey, or perhaps being pursued as prey.

This book offers an introduction to the complex world of tracking animals. Sometimes tracking is easy. At other times it is an incredible challenge that leaves you wondering just what animal made those unusual tracks. Take this book into the field with you, and it can provide some help with the first steps to identification. Animal tracks and trails are this book's focus; you will learn to recognize subtle differences for both. There are, of course, many additional signs to consider, such as scat and food caches, all of which help you to understand the animal you are tracking.

Remember, it takes many years to become an expert tracker. Tracking is one of those skills that grows with you as you acquire new knowledge in new situations. Most importantly, you will have an intimate experience with nature. You will learn the secrets of the seldom seen. The more you discover, the more you will want to know. By developing a good understanding of tracking, you will gain an excellent appreciation of the intricacies and delights of the marvelous natural world.

How to Use This Book

Most importantly, take this book into the field with you! Relying on your memory is not an adequate way to identify tracks. Track identification has to be done in the field, or with detailed sketches and notes that you can take home. Much of the process of identification involves circumstantial evidence, so you will have much more success when you are standing beside the track.

This book is laid out in an easy-to-use format. Beginning on p. 126, there is a quick reference appendix to the tracks of all the animals illustrated in the book. This appendix provides a fast way to familiarize yourself with certain tracks, and it guides you to the more informative descriptions of each animal and its tracks.

Each description is illustrated with the appropriate footprints and the track patterns that the animal usually leaves. Although these illustrations are not exhaustive, they do show the tracks or groups of prints that you will most likely see. You will find a list of dimensions for the tracks, giving the general range, but there will always be extremes, just as there are with people who have unusually small or large feet. Under the category 'Size' (of animal), the 'greater-than' sign (>) is used when the size difference between the sexes is pronounced.

If you think that you may have identified a track, check the 'Similar Species' section. This section is designed to help you confirm your conclusions by pointing out other animals that leave similar tracks and showing you ways to distinguish among them.

As you read this book, you will notice an abundance of words such as 'often,' 'mostly' and 'usually.' Unfortunately, tracking will never be an exact science; we cannot expect animals to conform to our expectations, so be prepared for the unpredictable.

Tips on Tracking

As you flip through this guide, you will notice clear, well-formed prints. Do not be deceived! It is a rare track that will ever show so clearly. For a good, clear print, the perfect conditions are slightly wet, shallow snow that isn't melting, or slightly soft mud that isn't actually wet. These conditions can be rare—most often you will be dealing with incomplete or faint prints, where you cannot even really be sure of the number of toes.

Should you find yourself looking at a clear print, then the job of identification is much easier. There are a number of key features to look for: measure the length and width of the print, count the number of toes, check for claw marks and note how far away

they are from the body of the print, and look for a heel mark. Keep in mind more subtle features, such as the spacing between the toes, if the toes are parallel and if fur on the sole of the foot has made the print less clear.

When you are faced with the challenge of identifying an unclear print—or even if you think that you have made a successful identification from one print alone—look beyond the single footprint and search out others. Do not rely on the dimensions of just one print, but collect measurements from several prints to get an average impression. Even the prints within one trail can show a lot of variation.

Try to determine which is the fore print and which is the hind, and remember that many animals are built very differently from humans, having larger forefeet than hind feet. Sometimes the prints will overlap, or they can be directly on top of one another in a direct register. For some animals, the fore and hind prints are pretty much the same.

Check out the pattern that the tracks make together in the trail and follow the trail for as many paces as is necessary for you to become familiar with the pattern. Patterns are very important and can be the distinguishing feature between different animals with otherwise similar tracks.

Follow the trail for some distance, because it can give you some vital clues. For example, the trail may lead you to a tree, indicating that the animal is a climber, or it may lead down into a burrow. This part of tracking can be the most rewarding, because you are following the life of the animal as it hunts, runs, walks, jumps, feeds or tries to escape a predator.

Take into consideration the habitat. Sometimes habitat alone will allow you to distinguish very similar tracks—one species might be found on riverbanks, whereas another might be encountered only in dense forest.

Think about your geographical location, too, because some animals have a limited range. This consideration can rule out some species and help you with your identification.

Remember that every animal will at some point leave a print or trail that looks just like the print or trail of a completely different animal!

Finally, keep in mind that if you track quietly, you might catch up with the maker of the prints.

Terms & Measurements

Some of the terms used in tracking can be rather confusing, and they often depend on personal interpretation. For example, what comes to your mind if you see the word 'hopping'? Perhaps you imagine a person hopping about on one leg—or perhaps you imagine a rabbit hopping through the countryside. Clearly, one person's perception of motion can be very different from another person's. Some useful terms are explained on the next few pages to clarify what is meant in this book, and, where appropriate, how the measurements given fit in with each term.

The following terms are sometimes used loosely and interchangeably—for example, a rabbit might be described 'a hopper' and a squirrel as 'a bounder,' yet both leave the same pattern of prints in the same sequence.

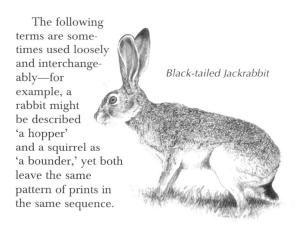

Black-tailed Jackrabbit

Ambling: Fast, rolling walking.

Bounding: A gait of four-legged animals in which the two hind feet land simultaneously, usually registering in front of the fore prints. It is common in rodents and the rabbit family. 'Hopping' or 'jumping' can often be substituted.

Gait: Describes how an animal is moving at some point in time. Different gaits result in different observable trail characteristics.

Galloping: A gait used by animals with four legs of even length, moving at high speed, hind feet registering in front of forefeet.

Hopping: Similar to bounding. With four-legged animals, it is usually indicated by tight clusters of prints, fore prints set between and behind the hind prints. A bird hopping on two feet creates a series of paired tracks along its trail.

Loping: Like galloping but slower, with each foot falling independently and leaving a trail pattern that consists of groups of tracks in the sequence fore-hind-fore-hind, usually roughly in a line.

Mustelids (weasel family) often use **2×2 loping**, in which the hind feet register directly on the fore prints. The resulting pattern has angled, paired tracks.

Running: Like galloping, but applied generally to animals moving at high speed. Also used for two-legged animals.

Trotting: Faster than walking, slower than running. The diagonally opposite limbs move simultaneously; that is, the right forefoot with the left hind, then the left forefoot with the right hind. This gait is the natural one for canids (dog family), short-tailed shrews and voles.

Canids may use **side-trotting**, a fast trotting in which the hind end of the animal shifts to one side. The resulting track pattern has paired tracks, with all the fore prints on one side and all the hind prints on the other.

Walking: A slow gait in which each foot moves independently of the others, resulting in an alternating track pattern. This gait is common for felines (cat family) and deer, as well as wide-bodied animals, such as bears and porcupines. Also used for two-legged animals.

Other Tracking Terms:

Dewclaws: Two small, toe-like structures set above and behind the main foot of most hoofed animals.

Direct Register: The hind foot falls directly on the fore print.

Double Register: The hind foot registers so as to overlap the fore print only slightly or falls beside it, so that both prints can be seen at least in part.

Dragline: A line left in snow or mud by a foot or the tail dragging over the surface.

Gallop Group: A track pattern of four prints made at a gallop, usually with the hind feet registering in front of the forefeet (see '***galloping***' for illustration).

Height: Taken at the animal's shoulder.

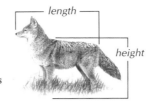

Length: The animal's body length from head to rump, not including the tail, unless otherwise indicated.

Metacarpal Pad: A small pad near the palm pad or between the palm pad and heel on the forefeet of bears and members of the weasel family.

Print (also called '***track***'): Fore and hind prints are treated individually. Print dimensions given are 'length' (including claws—maximum values may represent occasional heel register for some animals) and 'width.' A group of prints made by each of the animal's feet makes up a track pattern.

Register: To leave a mark—said about a foot, claw or other part of an animal's body.

Retractable: Describes claws that can be pulled in to keep them sharp, as with the cat family; these claws do not register in the prints. Foxes have semi-retractable claws.

Sitzmark: The mark left on the ground by an animal falling or jumping from a tree.

Straddle: The total width of the trail, all prints considered.

Stride: For consistency among different animals, the stride is taken as the distance from the center of one print (or print group) to the center of the next one. Some books may use the term 'pace.'

Track: Same as '***print***.'

Track Pattern: The pattern left after each foot registers once; a set of prints, such as a gallop group.

Trail: A series of track patterns; think of it as the path of the animal.

White-footed Mouse

MAMMALS

River Otter

White-tailed Deer

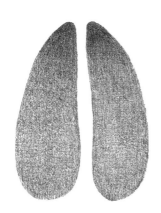

Fore and Hind Prints
Length: 2–3.5 in (5–9 cm)
Width: 1.6–2.5 in (4–6.5 cm)
Straddle
5–10 in (13–25 cm)
Stride
Walking: 10–20 in (25–50 cm)
Galloping: 6–15 ft (1.8–4.5 m)
Size (buck>doe)
Height: 3–3.5 ft (90–110 cm)
Length: to 6.3 ft (1.9 m)
Weight
120–350 lb (55–160 kg)

walking *gallop group*

WHITE-TAILED DEER
Odocoileus virginianus

The keen hearing of this deer guarantees that it knows about you before you know about it. Frequently, all that we see is its conspicuous white tail in the distance as it gallops away, earning this deer the nickname 'Flagtail.' The adaptable White-tailed Deer can be found throughout New Jersey in small groups at the edges of forests and in brushlands. It is common around ranches and residential areas.

This deer's prints are heart-shaped and pointed. Its alternating walking track pattern shows the hind prints direct registered or double registered on the fore prints. When a deer gallops on a soft surface, such as snow, the dewclaws register. This flighty deer gallops in the usual style, leaving hind prints ahead of fore prints, with the toes splayed wide for steadier, safer footing.

Similar Species: No other native cloven-hoofed mammal inhabits this state. In farming areas the prints of domesticated Elk (*Cervus elaphus*) or the Domestic Pig (*Sus scrofa*) can be mistaken for deer prints.

Horse

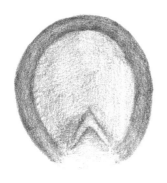

Fore Print
(hind print is slightly smaller)
Length: 4.5–6 in (11–15 cm)
Width: 4.5–5.5 in (11–14 cm)
Straddle
2–7.5 in (5–19 cm)
Stride
Walking: 17–28 in (43–70 cm)
Size
Height: to 6 ft (1.8 m)
Weight
to 1500 lb (680 kg)

walking

HORSE
Equus caballus

Outdoor adventures on horseback are a popular activity, so you can expect horse tracks to show up almost anywhere.

Unlike any other animal in this book, the Horse has just one huge toe on each foot. This toe leaves an oval print with a distinctive 'frog' (V-shaped mark) at its base. If a Horse is shod, the horseshoe shows up clearly as a firm wall at the outside of the print. Not all horses are shod, so do not expect to see this outer wall on every horse print. A typical, unhurried horse trail shows an alternating walking pattern, with the hind prints registered on or behind the slightly larger fore prints. Horses are capable of a range of speeds—up to a full gallop—but most recreational horseback riders take a more leisurely outlook on life, preferring to walk their horses and soak up the beautiful scenery!

Similar Species: Mules (rarely shod) have smaller tracks.

Black Bear

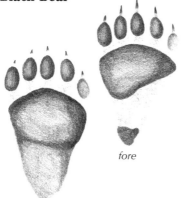

fore

hind

Fore Print
Length: 4–6.3 in (10–16 cm)
Width: 3.8–5.5 in (9.5–14 cm)
Hind Print
Length: 6–7 in (15–18 cm)
Width: 3.5–5.5 in (9–14 cm)
Straddle
9–15 in (23–38 cm)
Stride
Walking: 17–23 in (43–58 cm)
Size (male>female)
Height: 3–3.5 ft (90–110 cm)
Length: 5–6 ft (1.5–1.8 m)
Weight
200–600 lb (90–270 kg)

walking

BLACK BEAR
Ursus americanus

The Black Bear has a scattered range in forested areas of the eastern states, including New Jersey, but do not expect to see its tracks regularly. Habitat loss has severely decreased the number of Black Bears in the United States, and many regions now have only infrequent sightings. Finding fresh bear tracks can be a thrill, but take care—the bear may be just ahead. Never underestimate the potential power of a surprised bear!

Bear prints somewhat resemble small human prints, but wider and with claw marks. The small inner toe rarely registers. The forefoot's small heel pad often registers; the hind print shows a big heel. The bear's slow walk results in a slightly pigeon-toed double register with the hind print on the fore print. More frequently, at a faster pace, the hind foot oversteps the forefoot. When a bear runs, the extended clusters show the two hind feet in front of the forefeet. Along well-worn bear paths, look for 'digs' (patches of dug-up earth) and 'bear trees'—the scratched bark shows that this bear climbs.

Similar Species: No other wild animal is likely to leave similar tracks in this area.

Domestic Dog

fore

hind

Fore Print
(hind print is smaller)
Length: 1–5.5 in (2.5–14 cm)
Width: 1–5 in (2.5–13 cm)

Straddle
1.5–8 in (3.8–20 cm)

Stride
Walking: 3–32 in (7.5–80 cm)
Loping to Galloping: to 9 ft (2.7 m)

Size
Very variable

Weight
Very variable

walking   *loping to galloping*

DOMESTIC DOG
Canis familiaris

Dogs come in many shapes and sizes, from the tiny Chihuahua with its dainty feet to the robust and powerful Great Dane. Consequently, Domestic Dog tracks vary enormously. Dog ownership is high in many residential and rural areas, and where dogs are walked or allowed to roam free their tracks are left scattered about, especially if there is wet sand, mud or snow.

The forefeet of the Domestic Dog, which are much larger than the hind feet and support more of the animal's weight, leave the clearest tracks. When a dog walks, the hind prints usually register ahead of or beside the fore prints. As the dog moves faster, it trots and then lopes before it gallops. In a trot or lope pattern the prints alternate fore-hind-fore-hind, whereas a gallop group shows (from back to front) fore-fore-hind-hind.

Similar Species: Dog prints are usually found close to human tracks or activity. Fox (pp. 28–31) prints may be confused with small dog prints. Coyote (p. 26) prints are usually more oval and splay less.

Coyote

fore

hind

**Fore Print
(hind print is slightly smaller)**
Length: 2.4–3.2 in (6–8 cm)
Width: 1.6–2.4 in (4–6 cm)

Straddle
4–7 in (10–18 cm)

Stride
Walking: 8–16 in (20–40 cm)
Trotting: 17–23 in (43–58 cm)
Galloping/Leaping:
   2.5–10 ft (0.8 m–3 m)

Size (female is slightly smaller)
Height: 23–26 in (58–65 cm)
Length: 32–40 in (80–100 cm)

Weight
20–50 lb (9–23 kg)

walking or trotting   *gallop group*

COYOTE
(Brush Wolf, Prairie Wolf)
Canis latrans

This widespread, adaptable canine prefers open grasslands or woodlands. On its own, with a mate or in a family pack, it hunts rodents and larger prey. If you find a coyote den—usually a wide-mouthed tunnel leading into a nesting chamber—do not bother the family or the female will have to move her pups to a safer location.

The hind print is slightly smaller than the oval fore print, and its less-triangular heel pad rarely registers clearly. The claws of the two outer toes usually do not register. The Coyote typically walks or trots in an alternating pattern; the walk has a wider straddle, and the trotting trail is often very straight. When the Coyote gallops, its hind feet fall in front of its forefeet; the faster it goes, the straighter the gallop group. The Coyote's tail, which hangs down, leaves a dragline in deep snow.

Similar Species: A Domestic Dog's (p. 24) less-oval prints splay more, and its trail is erratic. Foot hairs blur Red Fox (p. 28) prints (usually smaller). Gray Fox (p. 30) prints are much smaller.

Red Fox

fore

hind

Fore Print
(hind print is slightly smaller)
Length: 2.1–3 in (5.3–7.5 cm)
Width: 1.6–2.3 in (4–5.8 cm)
Straddle
2–3.5 in (5–9 cm)
Stride
Trotting: 12–18 in (30–45 cm)
Side-trotting: 14–21 in (35–53 cm)
Size (vixen is slightly smaller)
Height: 14 in (35 cm)
Length: 22–25 in (55–65 cm)
Weight
7–15 lb (3.2–7 kg)

trotting   *side-trotting*

RED FOX
Vulpes vulpes

Very adaptable and intelligent, this beautiful and notoriously cunning fox is found throughout much of New Jersey, in a variety of habitats ranging from forests to open areas.

Abundant foot hair allows only parts of the toes and heel pads to register, but with no fine detail. The horizontal or slightly curved bar across the fore heel pad is diagnostic. A trotting Red Fox leaves a distinctive straight alternating trail—the hind print direct registers on the wider fore print. When the fox side-trots, its print pairs show the hind print to one side of the fore print in typical canid fashion. This fox gallops like the Coyote (p. 26). The faster the gallop, the straighter the gallop group.

Similar Species: Other canid prints lack the bar across the fore heel pad. Domestic Dog (p. 24) prints can be of similar size, but they will have a shorter stride and a less direct trail. Small Coyote prints are similar, but they have a wider straddle, and the toe marks are more bulbous. The Gray Fox (p. 30) makes smaller prints.

Gray Fox

fore

hind

**Fore Print
(hind print slightly smaller)**
Length: 1.3–2.1 in (3.3–5.3 cm)
Width: 1.1–1.5 in (2.8–3.8 cm)

Straddle
2–4 in (5–10 cm)

Stride
Walking/Trotting: 7–12 in (18–30 cm)

Size
Height: 14 in (35 cm)
Length: 21–30 in (53–75 cm)

Weight
7–15 lb (3.2–7 kg)

walking

GRAY FOX
Urocyon cinereoargenteus

This small, shy fox is widespread, but it especially prefers woodlands and chaparral country. The Gray Fox is the only fox that climbs trees, which it does either to seek safety or to forage.

The forefoot registers better than the smaller hind foot, and the hind foot's long, semi-retractable claws do not always register. The heel pads are often unclear—they sometimes show up just as small, round dots. When it walks, this fox leaves a neat alternating track pattern. When it trots, its prints fall in pairs, with the fore print set diagonally behind the hind print. The Gray Fox's gallop group is like the Coyote's (p. 26).

Similar Species: The Red Fox's (p. 28) fore heel pad has a bar across it; in general, Red Fox prints are larger and less clear (because of thick fur), and they have a longer stride and narrower straddle. Coyote tracks are much larger. Domestic Dog (p. 24) prints are much larger. Feline (pp. 32–35) prints lack claw marks, and they have larger, less symmetrical heel pads.

Bobcat

fore

hind

Fore Print
(hind print is slightly smaller)
Length: 1.8–2.5 in (4.5–6.5 cm)
Width: 1.8–2.5 in (4.5–6.5 cm)

Straddle
4–7 in (10–18 cm)

Stride
Walking: 8–16 in (20–40 cm)
Running: 4–8 ft (1.2–2.4 m)

Size (female is slightly smaller)
Height: 20–22 in (50–55 cm)
Length: 25–30 in (65–75 cm)

Weight
15–35 lb (7–16 kg)

walking   *ambling to loping*

BOBCAT
(Wildcat)
Lynx rufus

The handsome Bobcat, a stealthy and usually nocturnal hunter, is seldom seen. Very adaptable, it can leave tracks anywhere from wild mountainsides to chaparral and even into residential areas, but it is considered vulnerable and declining in New Jersey.

A walking Bobcat's hind feet usually register directly on its larger fore prints. As the Bobcat picks up speed, its trail becomes an ambling pattern of paired prints, the hind leading the fore. At even greater speeds it leaves four-print groups in a lope pattern. The fore prints, in particular, show asymmetry. The front part of the heel pad has two lobes and the rear part has three. The Bobcat's feet leave draglines in deep snow. Half-buried scat along the Bobcat's meandering trail marks its territory.

Similar Species: Large Domestic Cats (p. 34) make similar prints with a shorter stride and a narrower straddle. Four-toed mustelid (pp. 44–49) prints may also look similar. Canid (pp. 24–31) prints are narrower than they are long and show claw marks; the fronts of the footpads are once-lobed.

Domestic Cat

fore

hind

Fore Print
(hind print is slightly smaller)
Length: 1–1.6 in (2.5–4 cm)
Width: 1–1.8 in (2.5–4.5 cm)

Straddle
2.4–4.5 in (6–11 cm)

Stride
Walking: 5–8 in (13–20 cm)
Loping/Galloping:
    14–32 in (35–80 cm)

Size (male>female)
Height: 20–22 in (50–55 cm)
Length with tail: 30 in (75 cm)

Weight
6.5–13 lb (3–6 kg)

walking

loping to galloping

DOMESTIC CAT
(House Cat)
Felis catus

The tracks of the familiar, abundant Domestic Cat can show up almost any place where there are people. Abandoned cats may roam farther afield; these 'feral cats' lead a pretty wild and independent existence. Domestic Cats come in many shapes, sizes and colors.

As with all felines, a Domestic Cat's fore print and slightly smaller hind print both show four toe pads. Its retractable claws, kept clean and sharp for catching prey, do not register. Cat prints usually show a slight asymmetry, with one toe leading the others. A Domestic Cat makes a neat alternating walking track pattern, usually in direct register, as one would expect from this animal's fastidious nature. When a cat picks up speed, it leaves clusters of four prints, the hind feet registering in front of the forefeet.

Similar Species: A small Bobcat (p. 32) may leave tracks similar to a very large Domestic Cat's. Canid (pp. 24–31) prints will show claw marks. Four-toed mustelid (pp. 44–49) prints may also look similar.

Raccoon

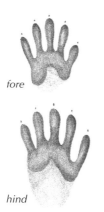

fore

hind

Fore Print
Length: 2–3 in (5–7.5 cm)
Width: 1.8–2.5 in (4.5–6.5 cm)
Hind Print
Length: 2.4–3.8 in (6–9.5 cm)
Width: 2–2.5 in (5–6.5 cm)
Straddle
3.3–6 in (8.5–15 cm)
Stride
Walking: 8–18 in (20–45 cm)
Bounding: 15–25 in (38–65 cm)
Size (female is slightly smaller)
Length: 24–37 in (60–95 cm)
Weight
11–35 lb (5–16 kg)

walking   *bounding group*

RACCOON
Procyon lotor

The inquisitive Raccoon is common throughout the region. Adored by some people for its distinctive face mask, it is disliked for its boundless curiosity—often demonstrated with residential garbage cans. Look for tracks near water at low elevations. The Raccoon likes to rest in trees. It usually dens up in cold weather.

The Raccoon's unusual print shows five well-formed toes. It looks like a human handprint; its small claws make dots. The highly dexterous forefeet rarely leave heel prints; the hind prints are generally much clearer and do show heels. The Raccoon's peculiar walking track pattern shows the left fore print next to the right hind print (or just in front) and vice versa. If it is in deep snow (which is rare), a Raccoon may use a direct-registering walk. It occasionally bounds, leaving clusters with the two hind prints in front of the fore prints.

Similar Species: Unclear Woodchuck (p. 66), Opossum (p. 38) or River Otter (p. 42) prints may look similar, but the latter two often drag their tails. The Muskrat (p. 64) makes smaller prints, and it often drags its tail.

Opossum

fore

hind

Fore Print
Length: 2–2.3 in (5–5.8 cm)
Width: 2–2.3 in (5–5.8 cm)
Hind Print
Length: 2.5–3 in (6.5–7.5 cm)
Width: 2–3 in (5–7.5 cm)
Straddle
4–5 in (10–13 cm)
Stride
5–11 in (13–28 cm)
Size
Length: 2–2.5 ft (60–75 cm)
Weight
9–13 lb (4–6 kg)

walking

fast walking

OPOSSUM
Didelphis virginiana

This slow-moving, nocturnal marsupial is found throughout New Jersey. It occupies many habitats and is quite tolerant of residential areas, but it prefers open woodland or brushland around waterbodies. Look for Opossum tracks in mud near the water and in snow during the warmer months of winter (the Opossum dens up during freezing weather) or near roadkill (which Opossums like to eat, though many suffer the same fate as the carrion that they dine on).

The Opossum has two walking habits: the common alternating pattern, with the hind prints registering on the fore prints, and a Raccoon-like (p. 36) paired-print pattern, with each hind print next to the opposing fore print. The very distinctive, long, inward-pointing thumb of the hind foot does not make a claw mark. The Opossum is an excellent climber. In snow, the long, naked tail's dragline may be bloodstained; not well adapted to cold, this thinly haired animal frequently suffers frostbite.

Similar Species: Prints in which the distinctive thumbs do not show may be mistaken for a Raccoon's.

Harbor Seal

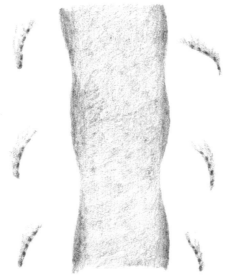

beach tracks

Size (male>female)
Length: 4–6 ft (1.2–1.8 m)
Weight
180–310 lb (80–140 kg)

HARBOR SEAL
Phoca vitulina

This seal is a migratory transient along the New Jersey coast, and it occasionally comes to shore. One of the smaller seals, it is quite shy and will usually slide off its rocky sentry post into the sea to escape curious naturalists. Look in sandy and muddy areas between platforms for its tracks—unmistakable because of their location and large size. On rare occasions, a Harbor Seal will work its way up a river, so you may find its tracks along riverbanks.

Seals, though unrivaled in the water, are not very graceful animals on land. A seal's heavy, fat body and flipper-like feet can leave messy tracks: a wide trough, (made by its cumbersome belly) with dents alongside (made as the seal pushed itself along with its forefeet). Look for the marks made by the seal's nails.

Similar Species: There are no other seals in the area. The unusual shape of the tracks makes confusion with the prints of other animals unlikely.

River Otter

fore

hind

Fore Print
Length: 2.5–3.5 in (6.5–9 cm)
Width: 2–3 in (5–7.5 cm)
Hind Print
Length: 3–4 in (7.5–10 cm)
Width: 2.3–3.3 in (5.8–8.5 cm)
Straddle
4–9 in (10–23 cm)
Stride
Loping: 12–27 in (30–70 cm)
Size
(female is two-thirds the size of male)
Length with tail: 3–4.3 ft (90–130 cm)
Weight
10–25 lb (4.5–11 kg)

loping (fast)

RIVER OTTER
Lontra canadensis

No animal knows how to
have more fun than a River Otter.
If you are lucky enough to watch one at play, you will
not soon forget the experience. Widespread and well-
adapted for the aquatic environment, this otter lives
near water; an otter in the forest is usually on its way to
another waterbody. The lithe River Otter loves to slide
in mud, wet grass or snow—often down riverbanks,
leaving troughs nearly 1 foot (30 cm) wide.

In soft mud, the River Otter's five-toed feet, especially the hind ones, register evidence of webbing. The
hind foot's inner toe is set slightly apart from the rest.
If the forefoot's metacarpal pad registers, it lengthens
the print. Very variable, otter trails usually show a
typical mustelid 2×2 loping, but with faster gaits they
show groups of four and three prints. The thick, heavy
tail often leaves a dragline.

Similar Species: Mink (p. 44) and Long-tailed Weasel
(p. 46) prints are about half the size, and do not have
a conspicuous tail dragline. Beavers (p. 62) are also
common around waterbodies, but the larger hind prints
will show more heel, longer toes and more webbing.

Mink

fore

hind

Fore and Hind Prints
Length: 1.3–2 in (3.3–5 cm)
Width: 1.3–1.8 in (3.3–4.5 cm)
Straddle
2.1–3.5 in (5.3–9 cm)
Stride
Walking/2×2 loping: 8–36 in (20–90 cm)
Size (male>female)
Length with tail: 19–28 in (48–70 cm)
Weight
1.5–3.5 lb (0.7–1.6 kg)

2×2 loping

MINK
Mustela vison

The lustrous Mink is widespread but uncommon in New Jersey. It prefers watery habitats surrounded by brush or forest. At home as much on land as in water, this nocturnal hunter can be exciting to track. Like the River Otter (p. 42), the Mink slides in mud and snow, carving out a trough up to 6 inches (15 cm) wide.

The Mink's fore print shows five (sometimes four) toes, with five loosely connected palm pads in an arc, but the hind print shows only four palm pads. The metacarpal pad of the forefoot rarely registers, but the furred heel of the hind foot may register, lengthening the hind print. The Mink prefers the typical mustelid 2×2 loping gait, which leaves consistently spaced, slightly angled double prints. Its diverse track patterns also include alternating walking, loping with three- and four-print groups (like the River Otter) and bounding (like a rabbit or hare, p. 52–57).

Similar Species: The Long-tailed Weasel (p. 46) makes similar but generally smaller tracks. The River Otter's trail shows distinct tail draglines. Bobcat (p. 32) prints may resemble four-toed Mink prints, but they will not show claw marks or lobed palm pads.

Long-tailed Weasel

Fore and Hind Prints
Length: 1.1–1.8 in (2.8–4.5 cm)
Width: 0.8–1 in (2–2.5 cm)

Straddle
1.8–2.8 in (4.5–7 cm)

Stride
Bounding: 9.5–43 in (24–110 cm)

Size (male>female)
Length with tail: 12–22 in (30–55 cm)

Weight
3–9.5 oz (85–270 g)

2x2 loping

LONG-TAILED WEASEL
Mustela frenata

The Long-tailed Weasel is the larger of New Jersey's two weasels. It is an active year-round hunter that has an avid appetite for rodents. It usually lives near water in open woodlands, brushy areas and some grasslands. Following this nimble creature's tracks can reveal much about its activities. Some weasel trails may lead you up a tree. Weasels sometimes take to water. Tracks are most evident in winter, when weasels frequently burrow into the snow or pursue rodents into their holes.

The usual weasel gait is a 2×2 lope, leaving a trail of paired prints. The Long-tailed Weasel's typical 2×2 lope shows an irregular stride—sometimes short and sometimes long—with no consistent behavior. Similar to the Mink (p. 44), this weasel may bound like a rabbit or hare (pp. 52–57).

Similar Species: Large Short-tailed Weasel (p. 48) tracks or small Mink tracks may be indistinguishable from Long-tailed Weasel tracks.

Short-tailed Weasel

Fore and Hind Prints
Length: 0.8–1.3 in (2–3.3 cm)
Width: 0.5–0.6 in (1.3–1.5 cm)
Straddle
1–2.1 in (2.5–5.3 cm)
Stride
2x2 loping: 9–36 in (23–90 cm)
Size (male>female)
Length with tail: 8–14 in (20–35 cm)
Weight
1.8–6 oz (45–170 g)

2x2 loping

SHORT-TAILED WEASEL
(Ermine, Stoat)
Mustela erminea

The Short-tailed Weasel is smaller than the Long-tailed Weasel (p. 46). It prefers woodlands and meadows up to higher elevations, but it does not favor wetlands or dense coniferous forests.

The result of this weasel's light weight, rapid movement and small, hairy feet is that the print detail is often unclear. Even with clear tracks, the inner (fifth) toe rarely registers. Although the typical weasel gait is a 2×2 lope that produces a trail of paired prints, this weasel's 2×2 loping tracks may fall in clusters, with alternating short and long strides.

Similar Species: Large Short-tailed Weasel tracks may be the same size as small Long-tailed Weasel tracks. Successfully distinguishing weasel tracks can be difficult. Close attention to the straddle and stride and track patterns is useful. Consideration of habitat may also help.

Striped Skunk

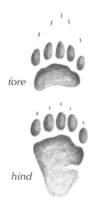

fore

hind

Fore Print
Length: 1.5–2.2 in (3.8–5.5 cm)
Width: 1–1.5 in (2.5–3.8 cm)

Hind Print
Length: 1.5–2.5 in (3.8–6.5 cm)
Width: 1–1.5 in (2.6–3.8 cm)

Straddle
2.8–4.5 in (7–11 cm)

Stride
Walking/Running:
   2.5–8 in (6.5–20 cm)

Size
Length with tail:
   20–32 in (50–80 cm)

Weight
6–14 lb (2.7–6.5 kg)

walking fast   *bounding*

STRIPED SKUNK
Mephitis mephitis

This striking skunk has a notorious reputation for its vile smell, and the lingering odor is often the best sign of its presence. Widespread throughout New Jersey, it prefers lower elevations in a diversity of habitats. The Striped Skunk dens up in winter, coming out on warmer days and in spring.

Both fore and hind feet have five toes. The long claws on the forefeet often register. The smooth palm pads and small heel pads leave surprisingly small prints. The Striped Skunk mostly walks—with such a potent smell for its defense, and those memorable black and white stripes, it rarely needs to run. Note that this skunk's trail rarely shows any consistent pattern, but an alternating walking pattern may be evident. The greater a skunk's speed, the more the hind foot oversteps the fore. If it runs, its trail consists of clumsy, closely set four-print groups. In snow it drags its feet.

Similar Species: Mink (p. 44) or Long-tailed Weasel (p. 46) tracks will be farther apart than those resulting from a skunk's shuffling gait. Skunk prints rarely overlap.

Black-tailed Jackrabbit

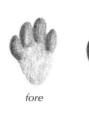

fore

hind

hopping

Fore Print
Length: 1.5–3 in (3.8–7.5 cm)
Width: 1.3–1.7 in (3.3–4.3 cm)
Hind Print
Length: 2.5–4 in (6.5–10 cm)
Length with heel: to 6 in (15 cm)
Width: 1.5–2.5 in (3.8–6.5 cm)
Straddle
4–7 in (10–18 cm)
Stride
Hopping: 5–10 ft (1.5–3 m)
Size
Length: 18–25 in (45–65 cm)
Weight
4–8 lb (1.8–3.6 kg)

BLACK-TAILED JACKRABBIT
Lepus californicus

Introduced to New Jersey, this athletic hare frequents open and agricultural areas, and it is sometimes found at higher elevations and in arid regions. Because of its nocturnal and solitary habits and its wariness of predators, it is infrequently seen.

Both fore and hind prints show four toes. The hind foot may register a long heel when the hare walks slowly. When it hops, this hare creates print groups in a triangular pattern; as it speeds up, these print groups spread out considerably. Following a hare's trail could lead you to its 'form'—a depression where it rests—or an urgent zigzag pattern that indicates where the hare fled from danger. With its strong hind legs, it is capable of leaping up to 20 feet (6 m) and running up to 35 mph (55 km/h) to avoid pursuers.

Similar Species: The European Hare (p. 54) makes very similar prints. The Eastern Cottontail (p. 56) makes a much smaller print cluster, and it has a shorter stride. Coyote (p. 26) prints resemble heel-less jackrabbit prints, but the pattern is very different.

European Hare

fore

hind

Fore Print
Length: 2–4 in (5–10 cm)
Width: 2 in (5 cm)
Hind Print
Length: 5 in (13 cm)
Length with heel: 9 in (23 cm)
Width: 2.5 in (6.5 cm)
Straddle
8 in (20 cm)
Stride
Hopping: 1.5–3 ft (45–90 cm)
Bounding: to 12 ft (3.7 m)
Size
Length: 25–30 in (65–75 cm)
Weight
7–12 lb (3.2–5.5 kg)

hopping

EUROPEAN HARE
Lepus europaeus

This large, sturdy colonizer of open fields was introduced to North America over one hundred years ago. It can now be found scattered throughout much of New Jersey. This rabbit easily adapts to many different habitats, and its numbers and range seem to be increasing. The European Hare is mainly brown, but it has a black spot on the top of its tail.

This hare's usual track pattern is typical for rabbits and hares: elongated triangular clusters of prints. The forefeet register first (usually in a slight diagonal line), leaving small, roundish prints that show four toes. The two much larger hind feet then fall (usually side by side) ahead of the fore prints. The hind prints are much longer when the whole heel registers, such as when the hare stops for a moment.

Similar Species: The smaller Eastern Cottontail (p. 56) has much smaller prints and a narrower straddle and stride. The Black-tailed Jackrabbit (p. 52) makes very similar prints.

Eastern Cottontail

fore

hind

hopping

Fore Print
Length: 1–1.5 in (2.5–3.8 cm)
Width: 0.8–1.3 in (2–3.3 cm)
Hind Print
Length: 3–3.5 in (7.5–9 cm)
Width: 1–1.5 in (2.5–3.8 cm)
Straddle
4–5 in (10–13 cm)
Stride
Hopping: 0.6–3 ft (18–90 cm)
Size
Length: 12–17 in (30–43 cm)
Weight
1.3–3 lb (0.6–1.4 kg)

EASTERN COTTONTAIL
Sylvilagus floridanus

This abundant rabbit can be found in much of New Jersey, although it seems to be more numerous in the south and the west. Preferring brushy areas in grasslands and cultivated areas, it might be found in dense vegetation, hiding from predators such as the Bobcat (p. 32) and the Coyote (p. 26). Largely nocturnal, the Eastern Cottontail might be seen at dawn or dusk and on darker days.

As with other rabbits and hares, this rabbit's most common track pattern is a triangular grouping of four prints, with the larger hind prints (which can appear pointed) falling in front of the fore prints (which may overlap). The hairiness of the toes will hide any pad detail. If you follow this rabbit's trail, you could be startled if it flies out from its 'form,' a depression in the ground in which it rests.

Similar Species: The New England Cottontail (*S. transitionalis*) makes identical prints. Both the European Hare (p. 54) and the Black-tailed Jackrabbit (p. 52) have much larger prints and a wider straddle and stride. Squirrel (pp. 70–77) tracks may show a similar pattern.

Porcupine

fore

hind

Fore Print
Length: 2.3–3.3 in (5.8–8.5 cm)
Width: 1.3–1.9 in (3.3–4.8 cm)
Hind Print
Length: 2.8–4 in (7–10 cm)
Width: 1.5–2 in (3.8–5 cm)
Straddle
5.5–9 in (14–23 cm)
Stride
Walking: 5–10 in (13–25 cm)
Size
Length with tail: 25–40 in (65–100 cm)
Weight
10–28 lb (4.5–13 kg)

walking

PORCUPINE
Erethizon dorsatum

This notorious rodent rarely runs—its many long quills are a formidable defense. Although its numbers are declining in New Jersey, you may encounter sign of this animal in forests and open woodlands.

The Porcupine's pigeon-toed, waddling gait leaves an alternating track pattern, with the hind print registered on or slightly in front of the shorter fore print. Look for long claw marks on all prints. The fore print shows four toes, and the hind print shows five. Clear prints may show the unusual pebbly surface of the solid heel pads, but a Porcupine's tracks are often scratch-marked by its heavy, spiny tail. In deeper snow this squat animal drags its feet, and it may leave a trough with its body. A Porcupine's trail might lead you to a tree, where this animal spends much of its time feeding; if so, look for chewed bark or nipped twigs on the ground.

Similar Species: The Porcupine's tracks are so distinctive that you are unlikely to confuse them with anything else.

Nutria

fore

hind

Fore Print
Length: to 3 in (7.5 cm)
Width: to 3 in (7.5 cm)
Hind Print
Length: 4.5–6 in (11–15 cm)
Width: to 3.5 in (9 cm)
Straddle
to 7 in (18 cm)
Stride
Walking: to 8 in (20 cm)
Size (male>female)
Length with tail:
 25–55 in (65–140 cm)
Weight
5–25 lb (2.3–11 kg)

walking

NUTRIA
Myocastor coypus

Much larger than a Muskrat (p. 64), this rodent was introduced from South America by fur farmers. It has escaped into the wild and formed numerous colonies in various states, including New Jersey. With its voracious appetite, the Nutria can be quite destructive to agricultural activities.

The webbing on the Nutria's strong hind feet is often evident in its prints. Look for claw marks too. The large hind print shows five toes, with one toe set farther back than the others. The much smaller fore print also shows five toes. The large, round, hairless tail often leaves a dragline. A Nutria's trail may lead you to a den in a riverbank, quite possibly a former Muskrat residence. Nearby, you might also notice large mats of vegetation—a Nutria's feeding platform.

Similar Species: Beaver (p. 62) prints are similar, but a Beaver's hind foot has more webbing (extending to the fifth toe). Muskrat prints are smaller.

Beaver

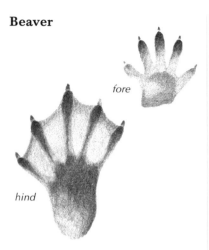

walking

Fore Print
Length: 2.5–4 in (6.5–10 cm)
Width: 2–3.5 in (5–9 cm)
Hind Print
Length: 5–7 in (13–18 cm)
Width: 3.3–5.3 in (8.5–13 cm)
Straddle
6–11 in (15–28 cm)
Stride
Walking: 3–6.5 in (7.5–17 cm)
Size
Length with tail: 3–4 ft (90–120 cm)
Weight
28–75 lb (13–34 kg)

BEAVER
Castor canadensis

Few animals leave as many signs of their presence as the Beaver, an animal capable of changing the local landscape. North America's largest rodent, it is a common sight around water. Look for its conspicuous dams and lodges and for the stumps of felled trees. Check trunks gnawed clean of bark for marks of the Beaver's huge incisors. Scent mounds marked with castoreum, a strong-smelling yellowish fluid that Beavers produce, also indicate recent activity.

Check the large hind prints for signs of webbing and broad toenails. It is rare for all five toes on each foot to register; often there is no sign of the nail of the fourth toe. Irregular foot placement in the alternating walking gait may produce a direct register or a double register. The Beaver's thick, scaly tail may mar its tracks, as can the branches that it drags about for construction and food. Repeated path use results in well-worn trails.

Similar Species: The Beaver's large hind prints and many other signs minimize confusion. Nutria (p. 60) prints are similar, but have less hind-foot webbing. Muskrat (p. 64) prints are much smaller.

Muskrat

fore

hind

Fore Print
Length: 1.1–1.5 in (2.8–3.8 cm)
Width: 1.1–1.5 in (2.8–3.8 cm)
Hind Print
Length: 1.6–3.2 in (4–8 cm)
Width: 1.5–2.1 in (3.8–5.3 cm)
Straddle
3–5 in (7.5–13 cm)
Stride
Walking: 3–5 in (7.5–13 cm)
Running: to 1 ft (30 cm)
Size
Length with tail: 16–25 in (40–65 cm)
Weight
2–4 lb (0.9–1.8 kg)

walking

MUSKRAT
Ondatra zibethicus

Like the Beaver (p. 62), this rodent can be found throughout New Jersey, wherever there is water. Beavers are very tolerant of Muskrats and even allow them to live in parts of their lodges. Active all year, the Muskrat leaves plenty of signs. It digs extensive networks of burrows, often undermining riverbanks, so do not be surprised if you suddenly fall into a hidden hole! Also look for small lodges in the water and beds of vegetation on which the Muskrat rests, suns and feeds in summer.

The small fifth (innermost) toe of the forefoot rarely registers. Stiff hairs that aid in swimming may create a 'shelf' around the five well-formed toes of the hind print. The common walking pattern shows print pairs that alternate from side to side; the hind print is just behind the fore print or slightly overlaps it. In snow, a Muskrat's feet and tail drag.

Similar Species: Few animals share this water-loving rodent's habits. The Beaver makes larger tracks, and it leaves many other signs as well. The Nutria (p. 60) makes larger prints.

Woodchuck

fore

hind

Fore and Hind Prints
Length: 1.8–2.8 in (4.5–7 cm)
Width: 1–2 in (2.5–5 cm)
Straddle
3.3–6 in (8.5–15 cm)
Stride
Walking: 2–6 in (5–15 cm)
Bounding: 6–14 in (15–35 cm)
Size (male>female)
Length with tail:
 20–25 in (50–65 cm)
Weight
5.5–12 lb (2.5–5.5 kg)

walking *bounding*

WOODCHUCK (Whistle Pig, Groundhog, Marmot)
Marmota monax

This robust member of the squirrel family is a common sight in open woodlands and adjacent open areas throughout New Jersey. Always on the watch for predators, but not too troubled by humans, the Woodchuck never wanders far from its burrow. This marmot hibernates during winter but emerges in early spring; look for its tracks in late spring snowfalls and in mud around burrow entrances.

A Woodchuck's fore print shows four toes, three palm pads and two heel pads (not always evident). The hind print shows five toes, four palm pads and two poorly registering heel pads. The Woodchuck usually leaves an alternating walking pattern, with the hind print registered on the fore print. When a Woodchuck runs from danger, it makes groups of four prints, hind ahead of fore.

Similar Species: Squirrel (pp. 70–77) prints are very similar but smaller. A small Raccoon's (p. 36) bounding track pattern will be similar, but it will show five-toed fore prints.

Eastern Chipmunk

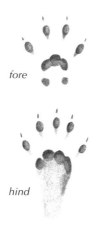

fore

hind

Fore Print
Length: 0.8–1 in (2–2.5 cm)
Width: 0.4–0.8 in (1–2 cm)
Hind Print
Length: 0.7–1.3 in (1.8–3.3 cm)
Width: 0.5–0.9 in (1.3–2.3 cm)
Straddle
2–3.2 in (5–8 cm)
Stride
Running: 7–15 in (18–38 cm)
Size
Length with tail: 7–10 in (18–25 cm)
Weight
2.5–5 oz (70–140 g)

bounding

EASTERN CHIPMUNK
Tamias striatus

Look for this delightful character throughout New Jersey. The Eastern Chipmunk is found in a variety of habitats, from the dense forest floor to open areas and even near buildings. You are more likely to see or hear this rodent, which is highly active during summer, than to notice its tracks. This large chipmunk is happiest on the ground, but it will gladly climb sturdy oak trees to harvest juicy, ripe acorns. It hibernates in winter, waking up from time to time to eat.

Chipmunks are so light that their tracks rarely show fine details. The forefeet each have four toes and the hind feet have five. Chipmunks run on their toes, so the two heel pads of the forefeet seldom register; the hind feet have no heel pads. Their erratic track patterns, like those of many of their cousins, show the hind feet registered in front of the forefeet. A chipmunk trail often leads to extensive burrows.

Similar Species: No other chipmunks live in this state. Squirrels (pp. 70–77) usually have larger prints and a wider straddle, and they are more likely to make midwinter tracks. Mouse (pp. 82–85) tracks are smaller.

Eastern Gray Squirrel

fore

hind

Fore Print
Length: 1–1.8 in (2.5–4.5 cm)
Width: 1 in (2.5 cm)
Hind Print
Length: 2.3–3 in (5.8–7.5 cm)
Width: 1.1–1.5 in (2.8–3.8 cm)
Straddle
3.8–6 in (9.5–15 cm)
Stride
Bounding: 0.7–3 ft (22–90 cm)
Size
Length with tail: 17–20 in (43–50 cm)
Weight
14–25 oz (400–710 g)

bounding

EASTERN GRAY SQUIRREL
Sciurus carolinensis

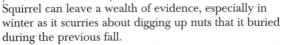

This large, familiar squirrel can be a common sight in deciduous and mixed forests throughout New Jersey, even in urban areas. Active all year, the Eastern Gray Squirrel can leave a wealth of evidence, especially in winter as it scurries about digging up nuts that it buried during the previous fall.

The Eastern Gray Squirrel leaves a typical squirrel track when it runs or bounds. The hind prints fall slightly in front of the fore prints. A clear fore print shows four toes with sharp claws, four fused palm pads and two heel pads. The hind print shows five toes and four palm pads; if the full heel-length registers, it also shows two small heel pads.

Similar Species: Fox Squirrel (p. 72) prints are as large or larger. Red Squirrel (p. 74) prints are smaller. Eastern Chipmunks (p. 68) and Southern Flying Squirrels (p. 76) make smaller tracks in a similar pattern, but they have narrower straddles. Rabbits and hares (pp. 52–57) make longer track patterns; their forefeet rarely register side by side.

Fox Squirrel

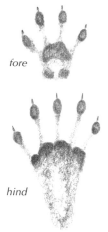

fore

hind

Fore Print
Length: 1–1.9 in (2.5–4.5 cm)
Width: 1–1.7 in (2.5–4.3 cm)
Hind Print
Length: 2–3.3 in (5–7.5 cm)
Width: 1.5–1.9 in (3.8–4.8 cm)
Straddle
4–6 in (10–15 cm)
Stride
Bounding: 0.7–3 ft (22–90 cm)
Size
Length with tail: 18–28 in (45–70 cm)
Weight
1–2.4 lb (0.5–1.1 kg)

bounding

FOX SQUIRREL
Sciurus niger

This squirrel is much like the Eastern Gray Squirrel (p. 70), but larger, and it has a yellowish underside. It can be a common sight in deciduous forests with plenty of nut trees and in open areas or woodlands. Piles of nutshells at tree bases indicate its favorite feeding sites. Active all year, the Fox Squirrel spends a lot of time foraging on the ground, often collecting nuts that it buried singly during the previous fall.

A clear fore print shows four toes with claws evident, four fused palm pads and two heel pads. The hind print shows five toes, four palm pads and sometimes a heel. When it runs or bounds, the Fox Squirrel makes a typical squirrel track, the hind prints slightly in front of the fore prints, with the prints in each pair roughly side by side.

Similar Species: Eastern Gray Squirrel prints are generally slightly smaller. Eastern Chipmunk (p. 68), Red Squirrel (p. 74) and Southern Fying Squirrel (p. 76) tracks fall in a similar pattern, but they are smaller, and the straddles are narrower. Rabbits and hares (pp. 52–57) make longer print groups, and their fore prints rarely register side by side.

Red Squirrel

fore

hind

Fore Print
Length: 0.8–1.5 in (2–3.8 cm)
Width: 0.5–1 in (1.3–2.5 cm)

Hind Print
Length: 1.5–2.3 in (3.8–5.8 cm)
Width: 0.8–1.3 in (2–3.3 cm)

Straddle
3–4.5 in (7.5–11 cm)

Stride
Bounding: 8–30 in (20–75 cm)

Size
Length with tail:
 9–15 in (23–38 cm)

Weight
2–9 oz (57–260 g)

bounding

bounding (deep snow)

RED SQUIRREL
(**Pine Squirrel, Chickaree**)
Tamiasciurus hudsonicus

When you enter the territory of a Red Squirrel, the inhabitant greets you with a loud, chattering call. Another obvious sign of this forest dweller, which is found throughout New Jersey, is its large middens—piles of cone scales and cores left beneath trees—that indicate favorite feeding sites.

Active all year in its small territory, a Red Squirrel will leave an abundance of trails that lead from tree to tree or down a burrow. This energetic animal mostly bounds, leaving groups of four prints, the hind prints in front of the fore prints, which tend to be side by side (but not always). Four toes show on each fore print, and five show on each hind print. The heels often do not register when squirrels move quickly. In deep snow the prints merge to form pairs of diamond-shaped tracks.

Similar Species: The larger Fox Squirrel (p. 72) and Eastern Gray Squirrel (p. 70) both make similar but larger tracks. Eastern Chipmunk (p. 68) and Southern Flying Squirrel (p. 76) tracks fall in a similar pattern, but they are smaller and have narrower straddles.

Southern Flying Squirrel

fore

hind

walking

Fore Print
Length: 0.3–0.5 in (0.8–1.3 cm)
Width: 0.4 in (1 cm)
Hind Print
Length: 0.9–1.3 in (2.3–3.3 cm)
Width: 0.5 in (1.4 cm)
Straddle
2–2.5 in (5–6.5 cm)
Stride
Bounding: 7–22 in (18–55 cm)
Size
Length with tail: 8–10 in (20–25 cm)
Weight
1.5–3.2 oz (43–90 g)

SOUTHERN FLYING SQUIRREL
Glaucomys volans

This soft-furred brown acrobat can glide long distances using the membranes that connect its forelegs and hindlegs. The largely nocturnal Southern Flying Squirrel lives in coniferous and mixed forests throughout the state. Up to 50 of these animals may huddle together in a nest in winter for warmth—they do not truly hibernate.

When it is on the ground, this flying squirrel normally bounds. Its hind prints may register in front of its fore prints, but New England–based tracking expert Mark Elbroch reports that its more common pattern (with a narrower straddle) shows the fore prints in front. If it glides down to the ground, this squirrel can leave a distinctive four-print 'sitzmark' in loose material or snow, but it often climbs down a tree trunk instead.

Similar Species: Other squirrels (pp. 70–75) usually make larger prints, and they rarely leave sitzmarks, but with unclear tracks it can be impossible to identify the species. Chipmunk (p. 68) tracks are very similar.

Norway Rat

fore

hind

Fore Print
Length: 0.7–0.8 in (1.8–2 cm)
Width: 0.5–0.7 in (1.3–1.8 cm)
Hind Print
Length: 1–1.3 in (2.5–3.3 cm)
Width: 0.8–1 in (2–2.5 cm)
Straddle
2–3 in (5–7.5 cm)
Stride
Walking: 1.5–3.5 in (3.8–9 cm)
Bounding: 9–20 in (23–50 cm)
Size
Length with tail: 13–19 in (33–48 cm)
Weight
7–18 oz (200–510 g)

walking

NORWAY RAT
(Brown Rat)
Rattus norvegicus

Active both day and night, this despised rat is widespread almost anywhere that humans have decided to build their homes. Not entirely dependent on people, it may live in the wild as well.

The fore print shows four toes, and the hind print shows five. When it bounds, this colonial rat leaves four-print groups, with the hind prints in front of the diagonally placed fore prints. Sometimes one of the hind feet direct registers on a fore print, creating a three-print group. This rat more commonly leaves an alternating walking pattern with the larger hind prints close to or overlapping the fore prints; the hind heel does not show. The tail often leaves a dragline in loose material. Rats live in groups, so you may find many trails together, often leading to their 5-inch (2-cm) wide burrows.

Similar Species: Mouse (pp. 82–85) prints are much smaller and show different track patterns. Chipmunks (p. 68) usually bound, and their tracks are often smaller. Squirrel (pp. 70–77) tracks show distinctive traits.

Woodland Vole

fore

hind

Fore Print
Length: 0.5 in (1.3 cm)
Width: 0.5 in (1.3 cm)
Hind Print
Length: 0.6 in (1.5 cm)
Width: 0.5–0.8 in (1.3–2 cm)
Straddle
1.3–2 in (3.3–5 cm)
Stride
Walking/Trotting: 0.8 in (2 cm)
Bounding: 2–6 in (5–15 cm)
Size
Length with tail:
 4–5.5 in (10–14 cm)
Weight
0.8–1.3 oz (23–37 g)

walking *bounding (in snow)*

WOODLAND VOLE (Pine Vole)
Microtus pinetorum

Distinguishing a vole track from those of the wealth of other small mammals in the state can be challenging. If you do spot a vole track, the Woodland Vole is a likely candidate. This adaptable rodent lives in a variety of habitats, not just woodlands.

When clear (which is seldom), vole fore prints show four toes, and hind prints show five. A vole's walk and trot both leave a paired alternating track pattern with a hind print occasionally direct registered on a fore print. Voles usually opt for a faster bounding gait in which the hind prints register on the fore prints to form print pairs. This vole lopes quickly across open areas, creating a three-print track pattern. Voles stay under the snow in winter; when it melts, look for distinctive piles of cut grass from their ground nests. The bark at the bases of shrubs may show tiny teeth marks left by gnawing. In summer, well-used vole paths appear as little runways in the grass.

Similar Species: The Southern Red-backed Vole (*Cletherionomys gapperi*) and the Meadow Vole (*M. pennsylvanicus*) are also common and leave indistinguishable tracks. Mouse (pp. 82–85) bounding tracks show four-print groups.

White-footed Mouse

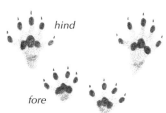

bounding group

Fore Print
Length: 0.3–0.4 in (0.8–1 cm)
Width: 0.3–0.4 in (0.8–1 cm)
Hind Print
Length: 0.3–0.5 in (0.8–1.3 cm)
Width: 0.3–0.4 in (0.8–1 cm)
Straddle
1.4–1.8 in (3.5–4.5 cm)
Stride
Bounding: 3–12 in (7.5–30 cm)
Size
Length with tail:
 6–12 in (15–30 cm)
Weight
0.5–1.3 oz (14–35 g)

bounding

bounding (in snow)

WHITE-FOOTED MOUSE
Peromyscus leucopus

The highly adaptable White-footed Mouse—one of the state's most abundant mammals—lives anywhere from woodlands to cultivated fields. It is seldom seen because it is nocturnal. This mouse may enter buildings in winter, where it will stay active. In colder areas it often hibernates.

In perfect, soft mud, the fore prints each show four toes, three palm pads and two heel pads, and the hind prints show five toes and three palm pads; the hind heel pads rarely register. Bounding tracks, most noticeable in snow, show the hind prints falling in front of the fore prints. In flaky snow the prints may merge to look like larger pairs of prints; tail drag is evident. This mouse's trail may lead up a tree or down into a burrow.

Similar Species: The Deer Mouse (*P. maniculatus*), which is rare in New Jersey, makes identical prints. The House Mouse (*Mus musculus*), with similar tracks, associates more with humans. Meadow Jumping Mouse (p. 84) prints show long, thin toes. Voles (p. 80) tend to trot, and they have a much shorter bounding group. Eastern Chipmunks (p. 68) have a wider straddle. Shrews (p. 86) have a narrower straddle.

Meadow Jumping Mouse

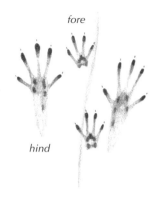

fore

hind

Fore Print
Length: 0.3–0.5 in (0.8–1.3 cm)
Width: 0.3–0.5 in (0.8–1.3 cm)

Hind Print
Length: 0.5–1.3 in (1.3–3.3 cm)
Width: 0.5–0.7 in (1.3–1.8 cm)

Straddle
1.8–1.9 in (4.5–4.8 cm)

Stride
Bounding: 7–18 in (18–45 cm)
In alarm: 3–6 ft (90–180 cm)

Size
Length with tail: 7–9 in (18–23 cm)

Weight
0.6–1.3 oz (17–35 g)

bounding

MEADOW JUMPING MOUSE
Zapus hudsonius

Congratulations if you find and successfully identify the tracks of the Meadow Jumping Mouse! Though it is abundant throughout the state, its preference for grassy meadows and dense undergrowth and its long, deep winter hibernation (about six months!) make locating tracks very difficult.

Jumping mouse tracks are distinctive if you do find them. The two smaller fore prints are between the long hind prints; the long heels do not always register, and some prints show just the three long middle toes. Occasionally, the toes on the forefeet splay so much that the side toes point backward. When they bound, jumping mice make short leaps. The tail may leave a dragline in soft mud or unseasonable snow. Clusters of cut grass stems about 5 inches (13 cm) long found lying in meadows are an abundant sign of this rodent.

Similar Species: White-footed Mouse (p. 82) tracks may have the same straddle. Heel-less jumping mouse hind prints may be mistaken for a vole's (p. 80), a small bird's (pp. 110) or even an amphibian's (pp. 114–19).

Masked Shrew

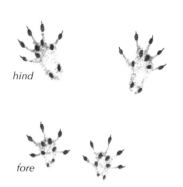

hind

fore

bounding group

Fore Print
Length: 0.2 in (0.5 cm)
Width: 0.2 in (0.5 cm)
Hind Print
Length: 0.6 in (1.5 cm)
Width: 0.3 in (0.8 cm)
Straddle
0.8–1.3 in (2–3.3 cm)
Stride
Bounding: 1.2–3 in (3–7.5 cm)
Size
Length with tail: 2.3–4.5 in (7–11 cm)
Weight
0.1–0.3 oz (3–9 g)

bounding

MASKED SHREW
Sorex cinereus

Though several species of tiny, frenetic shrews live in New Jersey, the widespread and adaptable Masked Shrew is a likely candidate if you find tracks. This small shrew prefers moist fields, marshes, bogs and woodlands, but it can also be found in higher and drier grasslands.

In its energetic and unending quest for food, a shrew usually leaves a four-print bounding pattern, but it may slow to an alternating walking pattern. The individual prints in a group are often indistinct but, in mud or shallow, wet snow, you can even count the five toes on each print. In deeper snow, a dragline is left by this shrew's tail. If a shrew tunnels under snow, it may leave a snow ridge on the surface. A shrew's trail may disappear down a burrow.

Similar Species: The Northern Short-tailed Shrew (*Blarina brevicauda*) makes very similar prints. Mouse (pp. 82–85) fore prints show four toes.

Star-nosed Mole

a molehill of the Star-nosed Mole

Size
Length with tail: 6–8.5 in (15–22 cm)
Weight
1–2.6 oz (28–75 g)

STAR-NOSED MOLE
Condylura cristata

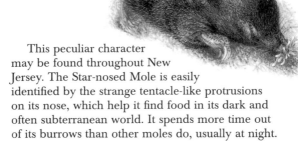

This peculiar character may be found throughout New Jersey. The Star-nosed Mole is easily identified by the strange tentacle-like protrusions on its nose, which help it find food in its dark and often subterranean world. It spends more time out of its burrows than other moles do, usually at night.

Typical signs of this mole include the big piles of soil pushed out of its burrows. Because of its preference for swimming and wet areas, look for this mole's hills along the banks of streams and rivers and in raised areas around marshes and wet fields. Clear prints from its long-clawed feet are rarely found. The Star-nosed Mole remains under the snow during winter and swims under ice, so even winter tracks are seldom seen.

Similar Species: The Hairy-tailed Mole (*Parascalops breweri*), which spends more time in its burrow, may leave similar signs, but usually not as close to water.

BIRDS, AMPHIBIANS & REPTILES

A guide to the animal tracks found in New Jersey is not complete without some consideration of the birds, amphibians and reptiles found in the state.

Several bird species have been chosen to represent the main types common to the state, but remember that individual bird species are not easily identified by tracks alone. Bird tracks are often abundant in snow and are clearest in shallow, wet snow. The shores of lakes and streams are very reliable places to find bird tracks—the mud there can hold a clear print for a long time. The sheer number of tracks made by shorebirds and waterfowl can be astonishing. Though some bird species prefer to perch in trees or soar across the sky, it can be entertaining to track birds that frequent the ground. Their tracks can spin around in circles and lead you in all directions. The trail may suddenly end as the bird takes flight, or it might terminate in a pile of feathers, the bird having fallen victim to a predator.

Many amphibians and turtles depend on moist environments, so look in the soft mud along the shores of lakes and ponds for their distinctive tracks. You may be able to distinguish frog tracks from toad tracks, because they generally move differently, but it can be very difficult to identify the species. Reptiles thrive and outnumber the amphibians in drier environments, but they seldom leave good tracks, except in occasional mud or perhaps in sand. Snakes leave distinctive body prints.

Mallard

Print
Length: 2–2.5 in (5–6.5 cm)
Straddle
4 in (10 cm)
Stride
to 4 in (10 cm)
Size
23 in (58 cm)

MALLARD
Anas platyrhynchos

male

female

This dabbling duck—the male a familiar sight with its striking green head—is common in open areas near lakes and ponds. Its webbed prints can often be seen in abundance along the muddy shores of just about any waterbody, including those in urban parks.

The webbed foot of the Mallard has three long toes that all point forward. Though the toes register well, the webbing between the toes does not always show in the print. The Mallard's inward-pointing feet give it a pigeon-toed appearance and perhaps account for its waddling gait, a characteristic for which ducks are known.

Similar Species: Many waterfowl, such as other ducks, as well as the Herring Gull (p. 94), leave similar prints. Exceptionally large prints were likely made by various species of geese.

Herring Gull

Print
Length: 3.5 in (9 cm)
Straddle
4–6 in (10–15 cm)
Stride
4.5 in (11 cm)
Size
Length: 23–25 in (58–65 cm)

HERRING GULL
Larus argentatus

The Herring Gull, with its long wings and webbed toes, is a strong long-distance flier as well as an excellent swimmer. This bird is increasingly common in New Jersey. It is concentrated in great numbers near waterbodies and garbage dumps.

Gulls leave slightly asymmetrical tracks that show three toes. They have claws that register outside the webbing, and the claw marks are usually attached to the footprint. Most gulls have quite a swagger to their gait, and they leave a trail with the tracks turned strongly inward.

Similar Species: Gull species cannot be reliably identified by track alone, but smaller species have conspicuously smaller tracks. Mallard (p. 92) and other duck tracks are often difficult to distinguish from gull tracks. Geese (various species) make larger prints.

Great Blue Heron

Print
Length: to 6.5 in (17 cm)
Straddle
8 in (20 cm)
Stride
9 in (23 cm)
Size
4.2–4.5 ft (1.3–1.4 m)

GREAT BLUE HERON
Ardea herodias

The regal and graceful image of this large heron symbolizes the precious wetlands in which it patiently hunts for food. Usually still and statuesque as it waits for a meal to swim by, this heron will have cause to walk from time to time, perhaps to find a better hunting location. Look for its large, slender tracks along the banks or mudflats of waterbodies.

Not surprisingly, a bird that lives and hunts with such precision walks in a similar fashion, leaving straight tracks that fall in a nearly straight line. Look for the slender rear toe in the print.

Similar Species: The American Bittern (*Botaurus lentiginosus*), the Green Heron (*Butorides virescens*) and the Black-crowned Night Heron (*Nyticorax nycticorax*) all make similar but smaller tracks.

Common Snipe

Print
Length: 1.5 in (3.8 cm)
Straddle
to 1.8 in (4.5 cm)
Stride
to 1.3 in (3.3 cm)
Size
11–12 in (28–30 cm)

COMMON SNIPE
Gallinago gallinago

This short-legged character is a resident of marshes and bogs, where its neat prints can often be seen in mud. Snipes are quite secretive when on the ground, and so you may be surprised if one suddenly flushes out from beneath your feet. If there is a Common Snipe in the air, you may hear an eerie whistle if it dives from the sky.

The Common Snipe's neat prints show four toes, including a small rear toe that points inward. The bird's short legs and stocky body give it a very short stride.

Similar Species: Many shorebirds, including the Spotted Sandpiper (p. 100), leave similar tracks.

Spotted Sandpiper

Print
Length: 0.8–1.3 in (2–3.3 cm)
Straddle
to 1.5 in (3.8 cm)
Stride
Erratic
Size
7–8 in (18–20 cm)

SPOTTED SANDPIPER
Actitis macularia

The bobbing tail of the Spotted Sandpiper is a common sight on the shores of lakes, rivers and streams, but you will usually find just one of these territorial birds in any given location. Because of its excellent camouflage, likely the first sign of this bird is it flying away, its fluttering wings close to the surface of the water.

As it teeters up and down on the shore, a sandpiper leaves trails of three-toed prints. Its fourth toe is very small and faces off to one side at an angle. Sandpiper tracks often have an erratic stride.

Similar Species: All sandpipers and plovers, including the common Killdeer (*Charadrius vociferus*), leave similar tracks, although there is much diversity in size. The Common Snipe (p. 98) makes similar but larger tracks.

Ruffed Grouse

Print
Length: 2–3 in (5–7.5 cm)
Straddle
2–3 in (5–7.5 cm)
Stride
Walking: 3–6 in (7.5–15 cm)
Size
15–19 in (38–48 cm)

RUFFED GROUSE
Bonasa umbellus

    This ground-dweller prefers the quiet seclusion of coniferous forests, where its excellent camouflage usually affords it good protection. Although you might find its tracks in mud, snow much improves your chances of finding them. If you follow a Ruffed Grouse trail quietly, you may be startled when the bird bursts from cover almost beneath your feet.

    The three thick front toes leave very clear impressions, but the short rear toe, which is angled off to one side, does not always show up so well. This bird's neat, straight trail appears to reflect its cautious approach to life on the forest floor.

Similar Species: No other grouse species live in New Jersey. The Wild Turkey (*Meleagris gallopavo*) leaves similar but much larger tracks.

Great Horned Owl

Strike
Width: to 3 ft (90 cm)
Size
22 in (55 cm)

GREAT HORNED OWL
Bubo virginianus

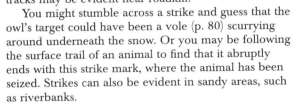

Often seen resting quietly in trees by day, this wide-ranging owl prefers to hunt at night. You might find an untidy hole in the snow, possibly surrounded by wing and tail-feather imprints—a well-registered 'strike' can be quite a sight. The Great Horned Owl strikes through the snow with its talons, and the feather imprints are made as the owl struggles to take off with possibly heavy prey. An ungraceful walker, it prefers to fly away from the scene, though its tracks may be evident near roadkill.

You might stumble across a strike and guess that the owl's target could have been a vole (p. 80) scurrying around underneath the snow. Or you may be following the surface trail of an animal to find that it abruptly ends with this strike mark, where the animal has been seized. Strikes can also be evident in sandy areas, such as riverbanks.

Similar Species: Several other birds might make a strike mark; strikes with less-rounded, more-distinct feather imprints might have been made by a hawk.

American Crow

Print
Length: 2.5–3 in (6.5–7.5 cm)
Straddle
1.5–3 in (3.8–7.5 cm)
Stride
Walking: 4 in (10 cm)
Size
16 in (40 cm)

AMERICAN CROW
Corvus brachyrhyncos

The black silhouette of the American Crow is a common sight in a variety of habitats. A crow will frequently come down to the ground and contentedly strut around. Its loud *caw* can be heard from quite a distance. Crows can be especially noisy when they are mobbing an owl or a hawk.

The American Crow typically leaves an alternating walking track pattern. Its prints show three sturdy toes pointing forward and one toe pointing backward. When a crow is in need of greater speed, perhaps for take-off, it bounds along leaving irregular pairs of diagonally placed prints with a longer stride between each pair.

Similar Species: The Fish Crow (*Corvus ossifragus*) makes slightly smaller prints. Other corvids also spend a lot of time on the ground and make similar tracks that vary in size according to the size of the bird.

Northern Flicker

Print
Length: 1.8 in (4.5 cm)
Straddle
1–1.5 in (2.5–3.8 cm)
Stride
Hopping: 1.5–5 in (3.8–13 cm)
Size
5.5–6.5 in (14–17 cm)

NORTHERN FLICKER
Colaptes auratus

This attractive woodpecker, which can be seen throughout New Jersey, is common in open woodlands and right into suburban areas. Unlike most woodpeckers, it spends some of its time feeding on the ground.

A clear flicker track shows a distinctive arrangement of two strong toes pointing forward and two pointing to the rear, with the outer toes slightly longer than the inner ones. The flicker's toes—along with short, strong legs that give the bird a short stride—are well suited for grasping tree trunks and limbs as this agile bird works its way along in search of insects.

Similar Species: Most other birds have very different tracks. Other woodpeckers leave similar tracks, but very few come down to the ground as much as the obliging Northern Flicker does.

Dark-eyed Junco

Print
Length: to 1.5 in (3.8 cm)
Straddle
1–1.5 in (2.5–3.8 cm)
Stride
Hopping: 1.5–5 in (3.8–13 cm)
Size
5.5–6.5 in (14–17 cm)

DARK-EYED JUNCO
Junco hyemalis

This common small bird typifies the many small hopping birds found in the region. Each foot has three forward-pointing toes and one longer toe at the rear. The best prints are left in mud or light snow. In loose material or deep snow the toe detail is lost and the footprints may show some dragging between the hops.

A good place to study this type of prints is near a birdfeeder. Watch the birds scurry around as they pick up fallen seeds, then have a look at the prints left behind. For example, juncos are attracted to small seeds that chickadees (*Poecile* spp.) scatter as they forage for sunflower seeds in the birdfeeder. Also look for tracks under coniferous trees, where juncos feed on fallen seeds in winter.

Similar Species: Note that the changing seasons influence the bird diversity in a region. Toe size may help with identification, because larger birds make larger prints. In powdery snow, junco tracks could be confused with mouse (pp. 82–85) tracks, so follow the trail to see if it disappears down a hole or into thin air.

Northern Cardinal

Print
Length to: 1.5 in (3.8 cm)
Straddle
1–1.5 in (2.5–3.8 cm)
Stride
Hopping: 1.5–5 in (3.8–13 cm)
Size
9 in (23 cm)

NORTHERN CARDINAL
Cardinalis cardinalis

female

male

The brilliant red plumage and small black mask and chin of the male are a joy to see on this year-round resident. Like the Dark-eyed Junco (p. 110), the Northern Cardinal is a small hopping bird.

Each foot has three forward-pointing toes and one longer toe at the rear. The best prints are left in snow, although in deep snow the toe detail is lost, and the feet may show some dragging between the hops.

A good place to study these types of prints is near a birdfeeder. Watch the birds scurry around as they pick up fallen seeds, then have a look at the prints that they have left behind.

Similar Species: Juncos, finches (various species) and sparrows (various species) make similar tracks. The size of the toes may indicate what kind of bird you are tracking—larger birds have larger footprints. Not all birds are present year-round, so keep in mind the season when tracking.

Frogs

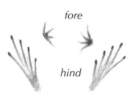

Straddle
to 3 in (7.5 cm)

hopping

FROGS

Wood Frog

The best place to look for frog tracks is along the muddy fringes of waterbodies. A frog's hopping action results in its two small forefeet registering in front of its long-toed hind prints. Frog tracks vary greatly in size, depending on species and age. Toads (p. 116) usually walk, but may also hop.

The smallest frogs include the treefrogs. The Spring Peeper (*Pseudacris crucifer*) is only 1.5 inches (3.8 cm) in length and prefers thick undergrowth and shrubs near the water, so its tracks a rare sight. The larger Gray Treefrogs (*Hyla chrysoscelis* and *H. versicolor*) spend most of their time in trees, coming down to breed and sing at night. The Green Frog (*Rana clamitans*) and the Pickerel Frog (*R. palustris*) both favor slow-moving, shallow water and swampy areas, and both are widespread. The beautiful and widespread Wood Frog (*R. sylvatica*), to 3 inches (8 cm) long, inhabits moist woodland areas. Unusually large tracks are surely from the robust Bullfrog (*R. catesbeiana*). Growing to 8 inches (20 cm) in length, it is North America's largest frog.

Toads

hind *fore*

Straddle:
to 2.5 in (6.5 cm)

walking

TOADS

American Toad

The best place to look for toad tracks is, as with frog tracks, undoubtedly along the muddy fringes of waterbodies. They can, however, occasionally be found in drier areas, as unclear trails in dusty patches of soil in woodlands, for example. In general, toads walk and frogs (p. 114) hop, but toads are pretty capable hoppers, too, especially when being hassled by overly enthusiastic naturalists. Toads leave rather abstract prints as they walk. The heels of the hind feet do not register. On less firm surfaces, the toes often leave draglines.

There are fewer toad species than frog species in New Jersey. The toad most likely to be encountered, and the most widespread, is the American Toad (*Bufo americanus*), which lives in many different moist habitats. Fowler's Toad (*B. fowleri*) is found scattered throughout the region in temporary pools and ditches. The Eastern Spadefoot (*Spea holbrooki*), found throughout New Jersey, is common in arid and semi-arid habitats. Toads in this region can be up to 5 inches (12 cm) in length.

Salamanders & Newts

Straddle:
to 3 in (7.5 cm)

walking

SALAMANDERS & NEWTS

Eastern Newt

There are several species of salamanders in New Jersey's moist and wet areas. Among the more abundant and widespread of these long, slender, lizard-like amphibians is the Eastern Newt (*Notophthalmus viridescens*); it can be up to 5.5 inches (14 cm) long. After a fresh rain, Eastern Newts emerging from ponds may leave small trails in the mud.

The Spotted Salamander (*Ambystoma maculatum*), which can grow to 10 inches (25 cm) long, lives throughout the region in mixed forests and some coniferous forests. The smaller Jefferson Salamander (*A. jeffersonianum*), with light blue spots on its undersides, grows to 8 inches (21 cm); it is found in northern New Jersey under debris near swamps and ponds in deciduous forests. The king of salamanders, the Eastern Tiger Salamander (*A. tigrinium*), is found in a variety of moist habitats, and it grows to 13 inches (33 cm) in length.

In general, a salamander's fore print shows four toes, and the larger hind print shows five. However, print detail is often blurred by the animal's dragging belly or by the swinging of its thick tail across the tracks.

Skinks

fore

hind

Straddle
to 3 in (7.5 cm)

walking

SKINKS

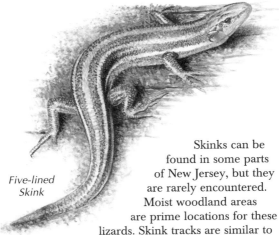

Five-lined Skink

Skinks can be found in some parts of New Jersey, but they are rarely encountered. Moist woodland areas are prime locations for these lizards. Skink tracks are similar to salamander tracks, but with longer, more slender toes.

If you find skink tracks, the most likely candidate is the Five-lined Skink (*Eumeces fasciatus*), which can grow to 8 inches (20 cm) long. It favors moist woodlands and is found throughout New Jersey. Another skink that you might encounter is the Ground Skink (*Scincella lateralis*), which grows up to 5.2 inches (13 cm) long. This handsome orange-brown skink has a distinct side stripe, and it inhabits humid forests.

These reptiles move very quickly when the need arises, their feet barely touching the ground as they dart for cover. Consequently, their tracks can be hard to make out clearly.

Turtles

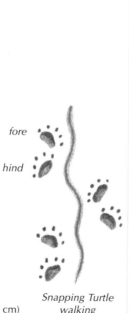

Straddle
4–10 in (10–25 cm)

Snapping Turtle walking

typical turtle walking

TURTLES

Painted Turtle

Turtles, those ancient inhabitants of the water world, will happily slip into the murky depths to avoid detection. They do, however, come out from time to time to feed or to bask in the sunshine. Look for their distinctive tracks alongside ponds, rivers and moist areas. Some turtles, such as the huge common Snapping Turtle (*Chelydra serpentina*) rarely come out of the water. One of the most widespread and prettiest turtles, often seen basking, is the Painted Turtle (*Chrysemys picta*); it can be almost 10 inches (25 cm) long. Slightly smaller is the well-named Common Musk Turtle (or Stinkpot, *Sternotherus odoratus*); it lives in shallow water and emits a foul odor. Smaller still is the distinctive Spotted Turtle (*Clemmys guttata*); it is common in beaver ponds. The Eastern Box Turtle (*Terrapene carolina*) is common in forested areas throughout the state.

With its large shell and short legs, a turtle leaves a track that is wide relative to the length of its stride—its straddle is about half its body length. Although longer-legged turtles can raise their shells off the ground, short-legged species may let them drag, as shown in their tracks. The tail may leave a straight dragline in the mud. On firmer surfaces, look for distinct claw marks.

Snakes

SNAKES

Common Garter Snake

Many snake species inhabit the state, and they are regularly encountered in a variety of habitats. Because snakes are all long and slender, their tracks appear so similar that accurate identification is next to impossible. In fact, because a snake lacks feet and leaves a track that is just a gentle meander, it is very challenging even to establish in which direction a snake was moving.

The most widespread and often encountered snake is the harmless Common Garter Snake (*Thamnophis sirtalis*). Found throughout the state, often close to wet or moist areas, it can be 4.3 feet (1.3 m) long. Also widespread is the Redbelly Snake (*Storeria occipitomaculata*), which prefers hilly woodlands and can reach 16 inches (40 cm) in length. The large Eastern Rat Snake (*Elaphe obsoleta*) inhabits forests, wooded canyons and grassy meadows along forest edges. This constrictor varies in color and can be up to 8.4 feet (2.5 m) long. The rattlesnake most likely to be found is the Timber Rattlesnake (*Crotalus horridus*), which can grow to be 6.3 feet (1.9 m) long; it frequents a variety of habitats from marshlands to dry woodlands. Also common in a variety of habitats is the beautiful Milk Snake (*Lampropeltis triangulum*).

TRACK PATTERNS & PRINTS

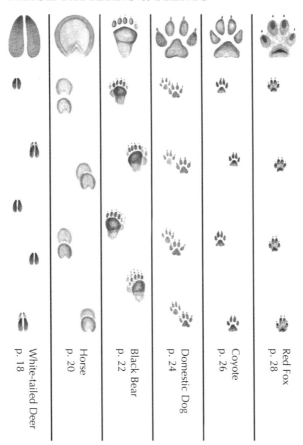

126

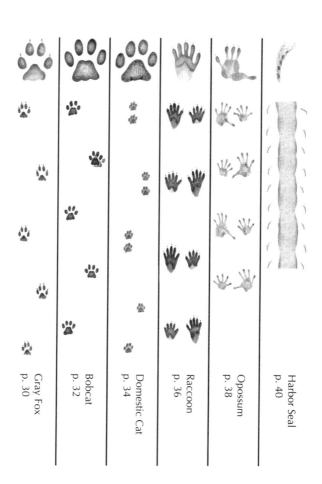

TRACK PATTERNS & PRINTS

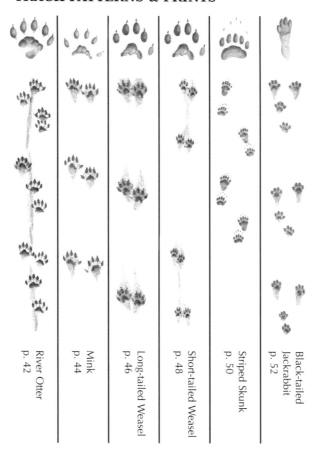

River Otter p. 42

Mink p. 44

Long-tailed Weasel p. 46

Short-tailed Weasel p. 48

Striped Skunk p. 50

Black-tailed Jackrabbit p. 52

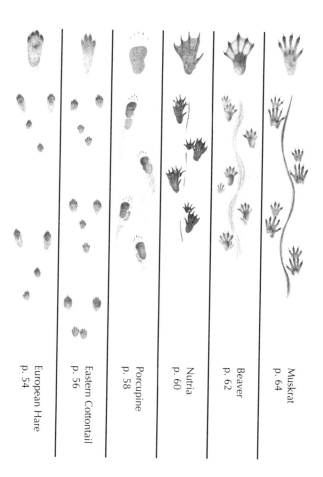

Muskrat
p. 64

Beaver
p. 62

Nutria
p. 60

Porcupine
p. 58

Eastern Cottontail
p. 56

European Hare
p. 54

TRACK PATTERNS & PRINTS

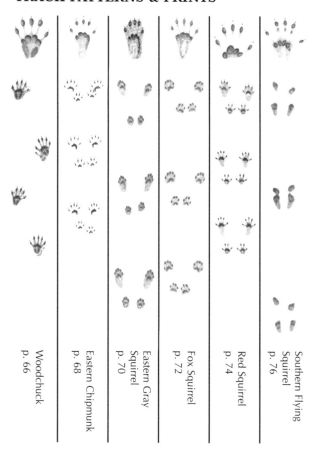

130

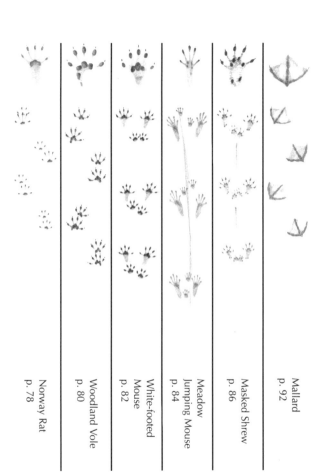

TRACK PATTERNS & PRINTS

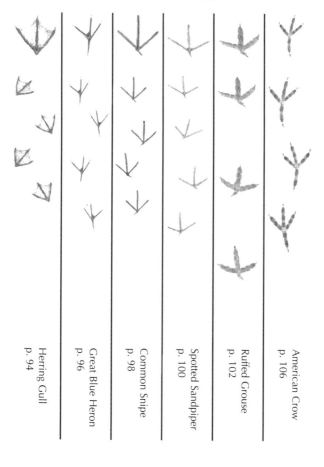

132

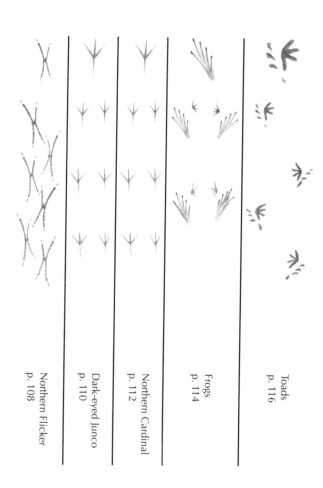

133

TRACK PATTERNS & PRINTS

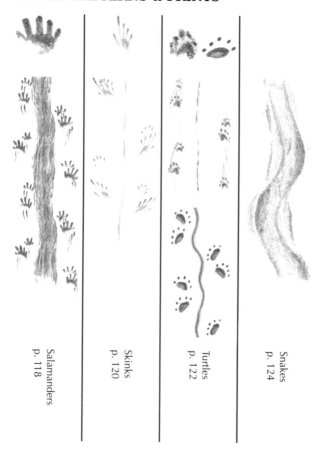

Salamanders p. 118

Skinks p. 120

Turtles p. 122

Snakes p. 124

HOOFED PRINTS

White-tailed Deer

Horse

HIND PRINTS

White-footed Mouse

Masked Shrew

Woodland Vole

Southern Flying Squirrel

Meadow Jumping Mouse

Eastern Chipmunk

135

HIND PRINTS

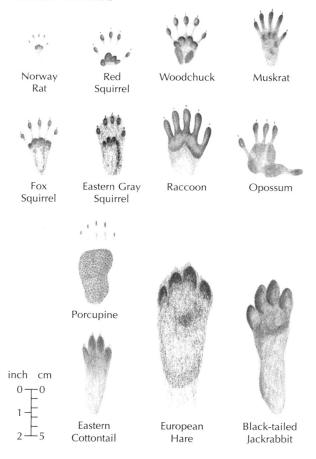

HIND PRINTS

Nutria

Beaver

Black Bear

inch cm
0 — 0
2
4 — 10

FORE PRINTS

Short-tailed Weasel

Long-tailed Weasel

Gray Fox

Red Fox

Domestic Cat

Mink

Striped Skunk

Coyote

Bobcat

River Otter

Domestic Dog

inch cm
0 — 0
1
2 — 5

BIBLIOGRAPHY

Behler, J.L., and F.W. King. 1979. *Field Guide to North American Reptiles and Amphibians*. National Audubon Society. New York: Alfred A. Knopf.

Brown, R., J. Ferguson, M. Lawrence and D. Lees. 1987. *Tracks and Signs of the Birds of Britain and Europe: An Identification Guide*. London: Christopher Helm.

Burt, W.H. 1976. *A Field Guide to the Mammals*. Boston: Houghton Mifflin Company.

Farrand, J., Jr. 1995. *Familiar Animal Tracks of North America*. National Audubon Society Pocket Guide. New York: Alfred A. Knopf.

Forrest, L.R. 1988. *Field Guide to Tracking Animals in Snow*. Harrisburg: Stackpole Books.

Halfpenny, J. 1986. *A Field Guide to Mammal Tracking in North America*. Boulder: Johnson Publishing Company.

Headstrom, R. 1971. *Identifying Animal Tracks*. Toronto: General Publishing Company.

Murie, O.J. 1974. *A Field Guide to Animal Tracks*. The Peterson Field Guide Series. Boston: Houghton Mifflin Company.

Rezendes, P. 1992. *Tracking and the Art of Seeing: How to Read Animal Tracks and Sign*. Vermont: Camden House Publishing.

Stall, C. 1989. *Animal Tracks of the Rocky Mountains*. Seattle: The Mountaineers.

Stokes, D., and L. Stokes. 1986. *A Guide to Animal Tracking and Behaviour*. Toronto: Little, Brown and Company.

Wassink, J.L. 1993. *Mammals of the Central Rockies*. Missoula: Mountain Press Publishing Company.

Whitaker, J.O., Jr. 1996. *National Audubon Society Field Guide to North American Mammals*. New York: Alfred A. Knopf.

INDEX

Page numbers in **boldface** type refer to the primary (illustrated) treatments of animal species and their tracks.

Actitis macularia, **100**
Ambystoma
  jeffersonianum, 119
  maculatum, 119
  tigrinium, 119
Amphibians, 85, 91, 114–19
Anas platyrhynchos, **92**
Ardea herodias, **96**

Bear, Black, **22**
Beaver, 43, 61, **62**, 65
Birds, 85, 91–113
Bittern, American, 97
Blarina brevicauda, 87
Bobcat, **32**, 35, 45, 57
Bonasa umbellus, **102**
Botaurus lentiginosus, 97
Bubo virginianus, **104**
Bufo
  americanus, 117
  fowleri, 117
Bullfrog, 115
Butorides virescens, 97

Canids, 24–31, 33, 35
Canis
  familiaris, **24**
  latrans, **26**
Cardinal, Northern, **112**
Cardinalis cardinalis, **112**
Castor canadensis, **62**
Cat
  Domestic, 33, **34**
  Feral. *See* Domestic C.
  House. *See* Domestic C.
Cervus elaphus, 19
Charadrius vociferus, 101
Chelydra serpentina, 123
Chickadees, 111
Chickaree. *See* Squirrel, Red
Chipmunk, Eastern, **68**, 71, 73, 75, 77, 79, 83
Chrysemys picta, 123
Clemmys guttata, 123
Cletherionomys gapperi, 81
Colaptes auratus, **108**
Condylura cristata, **88**

Corvus
  brachyrhyncos, **106**
  ossifragus, 107
Cottontail. *See also* Rabbits
  Eastern, 53, 55, **56**
  New England, 57
Coyote, 25, **26**, 29, 31, 53, 57
Crotalus horridus, 125
Crow
  American, **106**
  Fish, 107

Deer, White-tailed, **18**
Didelphis virginiana, **38**
Dog, Domestic, **24**, 27, 29, 31

Elaphe obsoleta, 125
Elk, 19
Equus caballus, **20**
Erethizon dorsatum, **58**
Ermine.
  See Weasel, Short-tailed
Eumeces fasciatus, 121

Felines, 31, 32–35
Felis catus, **34**
Finches, 113

Flagtail.
  See Deer, White-tailed
Flicker, Northern, **108**
Fox, 25
  Gray, 27, 29, **30**
  Red, 27, **28**, 31
Frog, 91. *See also* Bullfrog; Peeper; Treefrog
  Green, 115
  Pickerel, 115
  Wood, 115

Gallinago gallinago, **98**
Geese, 93, 95
Glaucomys volans, **76**
Groundhog. *See* Woodchuck
Grouse, Ruffed, **102**
Gull, Herring, 93, **94**

Hare, 45, 47, 71, 73. *See also* Jackrabbit
  European, 53, **54**, 57
Hawks, 105
Heron
  Black-crowned Night, 97
  Great Blue, **96**
  Green, 97
Horse, **20**

Hyla
  chrysoscelis, 115
  versicolor, 115

Jackrabbit. *See also* Hare
  Black-tailed, **52**, 55
Jumping Mouse, 83
  Meadow, **84**
Junco, Dark-eyed, **110**, 113
Junco hyemalis, **110**

Killdeer, 101

Lampropeltis
  triangulum, 125
Larus argentatus, **94**
Lepus
  californicus, **52**
  europaeus, 54
Lizard. *See* Skink
Lontra canadensis, **42**
Lynx rufus, **32**

Mallard, **92**, 95
Marmot. *See* Woodchuck
Marmota monax, **66**
Meleagris gallopavo, 103
Mephitis mephitis, **50**
Microtus
  pennsylvanicus, 81
  pinetorum, **80**

Mink, 43, **44**, 47, 51
Mole
  Hairy-tailed, 89
  Star-nosed, **88**
Mouse, 69, 79, 81, 87, 111.
  See also Jumping Mouse
  Deer, 83
  House, 83
  White-footed, **82**, 85
Mule, 21
Mus musculus, 83
Muskrat, 37, 61, 63, **64**
Mustela
  erminea, **48**
  frenata, **46**
  vison, **44**
Mustelids, 33, 35, 42–49
Myocastor coypus, **60**

Newt, Eastern, 119
Notophthalmus
  viridescens, 119
Nutria, **60**, 63, 65
Nyticorax nycticorax, 97

Odocoileus virginianus, **18**
Ondatra zibethicus, **64**
Opossum, 37, **38**
Otter, River, 37, **42**, 45
Owl, Great Horned, **104**

Parascalops breweri, 89
Peeper, Spring, 115
Peromyscus
  leucopus, **82**
  maniculatus, 83
Phoca vitulina, **40**
Pig, Domestic, 19
Plovers, 101
Poecile spp., 111
Porcupine, **58**
Procyon lotor, **36**
Pseudacris crucifer, 115

Rabbit, 45, 47, 71, 73.
  See also Cottontails
Raccoon, **36**, 39, 67
Rana
  catesbeiana, 115
  clamitans, 115
  palustris, 115
  sylvatica, 115
Rat
  Brown. *See* Norway R.
  Norway, **78**
Rattlesnake, Timber, 125
Rattus norvegicus, **78**
Reptiles, 91, 120–25

Salamander
  Eastern Tiger, 119
  Jefferson, 119
  Spotted, 119

Sandpiper, Spotted,
  99, **100**
Scincella lateralis, 121
Sciurus
  carolinensis, **70**
  niger, **72**
Seal, Harbor, **40**
Shrew, 83
  Masked, **86**
  Northern Short-tailed, 87
Skink
  Five-lined, 121
  Ground, 121
Skunk, Striped, **50**
Snake, 91. *See also*
  Rattlesnake, Timber
  Common Garter, 125
  Eastern Rat, 125
  Milk, 125
  Redbelly, 125
Snipe, Common, **98**, 101
Sorex cinereus, **86**
Spadefoot, Eastern, 117
Sparrows, 113
Spea holbrooki, 117
Squirrel, 57, 67, 69, 79
  Eastern Gray, **70**, 73, 75
  Fox, 71, **72**, 75
  Pine. *See* Red S.
  Red, 71, 73, **74**
  Southern Flying,
    71, 73, 75, **76**

Sternotherus odoratus, 123
Stinkpot. *See* Turtle,
   Common Musk
Stoat. *See* Weasel, Short-tailed
Storeria occipitomaculata, 125
Sus scrofa, 19
Sylvilagus
  floridanus, **56**
  transitionalis, 57

Tamias striatus, **68**
Tamiasciurus hudsonicus, **74**
Terrapene carolina, 123
Thamnophis sirtalis, 125
Toad, 91, 115. *See also*
   Spadefoot
  American, 117
  Fowler's, 117
Treefrog, Gray, 115
Turkey, Wild, 103
Turtle, 91
  Common Musk, 123
  Eastern Box, 123
  Painted, 123
  Snapping, 123
  Spotted, 123

Urocyon cinereoargenteus, **30**
Ursus americanus, **22**

Vole, 83, 85, 105
  Meadow, 81
  Pine. *See* Woodland V.
  Southern Red-backed, 81
  Woodland, **80**
Vulpes vulpes, **28**

Weasel
  Long-tailed, 43, 45,
    46, 49, 51
  Short-tailed, 47, **48**
Whistle Pig. *See* Woodchuck
Wildcat. *See* Bobcat
Wolf
  Brush. *See* Coyote
  Prairie. *See* Coyote
Woodchuck, 37, **66**
Woodpeckers, 109

Zapus hudsonius, **84**

ABOUT THE AUTHORS

Tamara Eder, equipped from the age of six with a canoe, a dip net and a note pad, grew up with a fascination for nature and the diversity of life. She has a degree in environmental conservation sciences and has photographed and written about the biodiversity in Bermuda, the Galapagos Islands, the Amazon Basin, China, Tibet, Vietnam, Thailand and Malaysia. Tamara is also the author of Lone Pine's Mammal series and Whales series.

Ian Sheldon, an accomplished artist, naturalist and educator, has lived in South Africa, England and Singapore. Living in different countries with different ecosystems encouraged Ian's desire to study mammals, birds and other creatures, and he earned an award from the Zoological Society of London and a degree from Cambridge University. He has also completed a Master's degree in Ecotourism Development. As an artist, Ian is represented by galleries internationally, and he is both a writer and illustrator of many other nature guides, including Lone Pine's *Seashore of Northern and Central California*, *Animal Tracks of New England* and the upcoming *Bugs of Washington and Oregon*.